PROJECT M²: MENTORING YOUNG MATHEMATICIANS

Shopping at the Bazaar

Connecting Number and Algebra with the Meerkats

M. Katherine Gavin, *University of Connecticut*
Tutita M. Casa, *University of Connecticut*
Suzanne H. Chapin, *Boston University*
Linda Jensen Sheffield, *Northern Kentucky University*

Student Mathematician's Journal

Connections to the Common Core State Standards

Kendall Hunt
publishing company

Meerkat photo on pg 25: © Pushish Images/Shutterstock.com

Kendall Hunt
publishing company

www.kendallhunt.com
Send all inquiries to:
4050 Westmark Drive
Dubuque, IA 52004-1840
1-800-542-6657

ISBN 978-1-5249-1002-0

Published in the United States of America

17 16 15 14 13

Production Date: 2017
Printed by: LSI
United States of America
Batch number: 441002

Table of Contents

Table of Contents

As **Speakers** we will:

1. Talk loud enough for others to hear.

2. Turn to talk to the class.

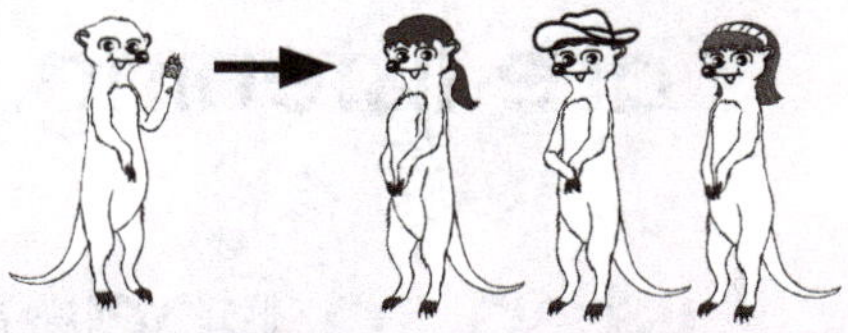

3. Share different ideas.

4. Explain our ideas.

5. Agree and disagree with ideas, not with each other.

As **Listeners** we will:

1. Ask speakers to speak up.
2. Show speakers we are listening.

3. Listen to understand.
4. Ask questions to make sense of the idea.

Did you mean...?

5. Think carefully about all speakers' ideas.

As **Writers** we will:

Think — Think about the question.

Talk — Talk about the answer.

Tell — Tell:

1. all ideas
2. the answer
3. "why"

Thinking Like a Mathematician*

Here is a list of skills mathematicians use every day. See how many you can use in your Student Mathematician's Journal.

1. Work hard to understand and solve problems.
2. Make sense of numbers in problems and use operations with numbers.
3. Defend your math ideas. Explain why you agree or disagree with someone's thinking.
4. Use the math you know to help solve problems in your world.
5. Choose and use the right math tools to help solve problems.
6. Use correct math vocabulary and symbols to explain clearly.
7. Look for and use patterns to help solve problems.
8. Notice if you are using the same math again and again and look for short cuts.
9. Solve a problem in a new way. Ask new questions to investigate.**

*Adapted from the Common Core State Standards: Standards for Mathematical Practice

National Governors Association Center for Best Practices (NGA Center), Council of Chief State School Officers (CCSSO). (2010). *Common core state standards for mathematics.* Washington, DC: Authors. Retrieved from http://www.corestandards.org/the-standards.

** Johnsen, S. K., & Sheffield, L. J. (Eds.). (2013). *Using the common core state standards for mathematics with gifted and advanced learners.* Waco, TX : Prufrock Press.

Student Mathematician: ______________________ Date: __________

Our Agate Stone Designs

Record your designs for Dru and Teller here!

Student Mathematician: ______________________ Date: __________

THINK DEEPLY

Dear Dru and Teller,

We think that there are a lot of different combinations of agate stones that you can sell to the *Twice as Nice Belt Shop* owners. One reason we know that all of our pairs of bags are balanced is because

____

____

A second reason is because ______________________

____

____

Our drawings show you some of our designs. On each page, one drawing shows you how we put stones in the bags and the other one shows you how we took some stones out of at least one bag. We hope you like our ideas!

Sincerely,

____

Student Mathematician: ______________________ Date: __________

THINK DEEPLY

Here is one idea for how to put agate stones in two bags.

Here is one idea for how to take out agate stones from at least one bag.

Student Mathematician: ______________________ Date: ____________

True or False?

1. Write a “T” if you think it is true or “F” if you think it is false.
2. Compare with your partner. Talk about why you agree or disagree.
3. If you agree the number sentence is false, put a “/” through the equal sign to show the expressions are not equal (≠).

Partner 1	Number Sentence	Partner 2
	a) $14 + 5 = 19$	
	b) $30 = 28 - 2$	
	c) $16 + 4 = 20 + 2$	
	d) $143 = 143$	
	e) $36 - 14 = 36 - 14$	
	f) $19 - 3 = 14 + 2$	
	g) $153 - 26 = 152 - 26$	
	h) $216 + 72 = 72 + 216$	
	i) $5 + 3 = 16 - 8$	
	j) $62 + 5 = 67 + 2$	
	k) $78 - 0 = 78$	
	l) $0 + 125 = 125$	
	m) $45 + 30 = 100 - 25$	

4. Explain why you both think $16 + 4 = 20 + 2$ (number sentence c) is true or false.

5. Explain why you both think $19 - 3 = 14 + 2$ (number sentence f) is true or false.

Student Mathematician: ______________________________ Date: ____________

Izit? Recording Sheet

Round	Number Sentence	True or False?
1		
2		
3		
4		
5		
6		
7		
8		
9		
10		

We built ____________________ true number sentences or equations.

We built ____________________ false number sentences.

Student Mathematician: ______________________ Date: ____________

THINK DEEPLY

We have 37 agates in one bag and 35 of them in a second bag. Please tell us different ways we can make the same number of agates in each bag!

Use words and symbols to show us.

Dear Dru and Teller,

Student Mathematician: ________________________ Date: ____________

Break It Up Recording Sheet

Record your *Break It Up* equations below.

EXAMPLE: **Show how you BREAK IT UP and REMAKE IT!**

$18 + 59$

$17 + 1 + 59$

$17 + 60 = 77$

1.

2.

3.

4.

5.

6.

Student Mathematician: ______________________ Date: __________

Thinking About Addition

Find the missing values to make equations.

1. $63 + 83 = 84 + \square$

2. $57 + 29 = \square + 56$

3. $\square + 317 = 84 + 315$

4. $124 + 79 = 76 + \square$

5. $49 + \square = 214 + 51$

6. $\square + 722 = 718 + 820$

7. $\square + 2{,}354 = 191 + 2{,}350$

8. $478 + 69 = 63 + \square$

Explain your thinking to find the value of $\square$ in Equation 8.

____

____

____

____

Student Mathematician: ______________________ Date: __________

THINK DEEPLY

a. Explain to Dru and Teller how Suma can easily figure out the number of agate stones to put in the bag when given the following equation:

$$246 + \square = 319 + 248$$

b. Use words, numbers, pictures, or diagrams to show that both expressions are equal.

____

____

____

____

____

____

____

____

____

Student Mathematician: ______________________ Date: ____________

Number Lines and Subtraction

Use these number lines to write five subtractions with a difference of 4. Use arrows or paper strips to help you.

Example:

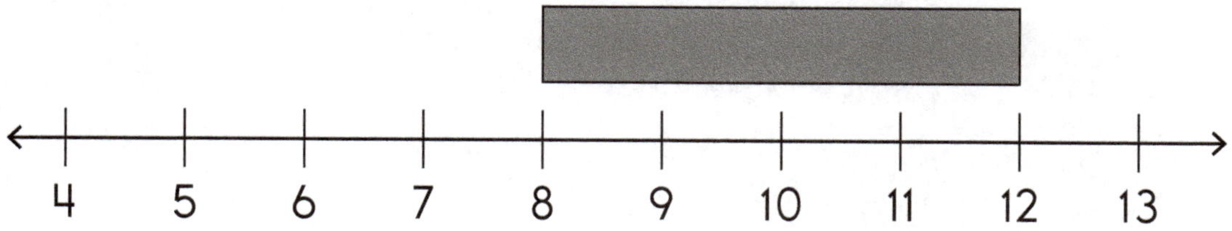

$$\begin{array}{r} 12 \\ -8 \\ \hline 4 \end{array}$$

1.

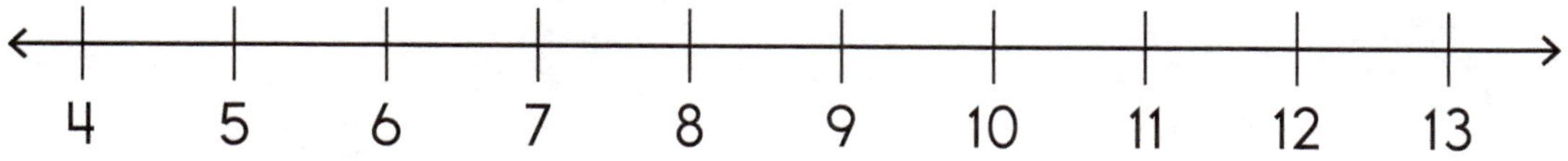

2.

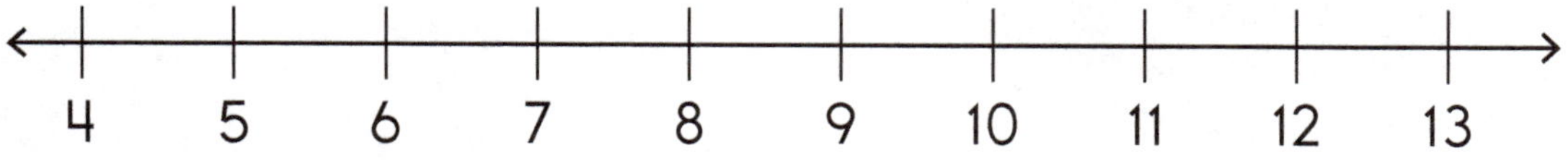

3.

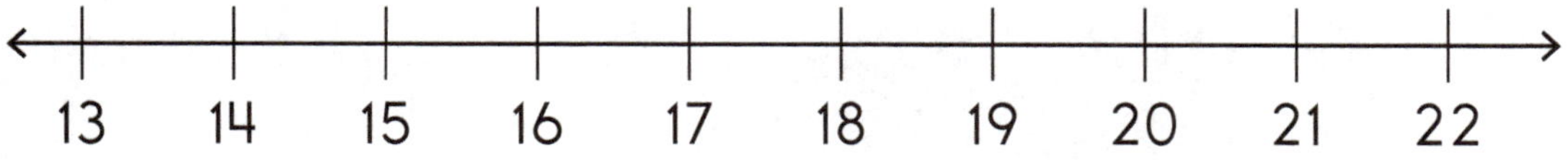

4.

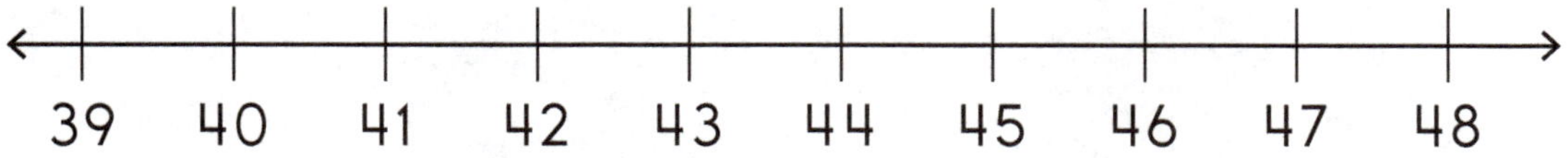

5.

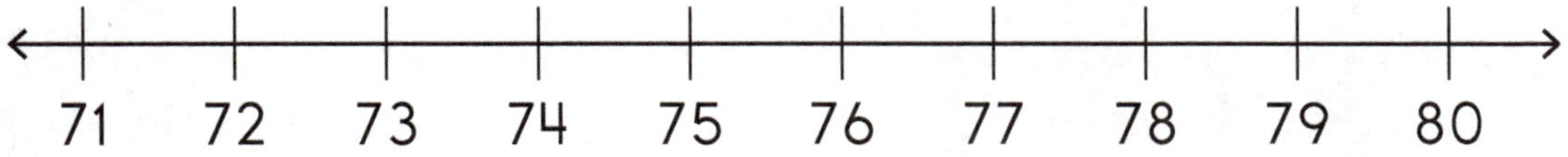

Student Mathematician: ______________________________ Date: ______________

Mysterious Mula Subtractions

Dru and Teller bought some beads in the market for 17 mula. They had 52 mula to pay the shopkeeper. The shopkeeper needed to give the Duo some change. Here's what the shopkeeper said: "52 − 17 = 55 − 20. That's easier!"

1. What number did the shopkeeper add to the minuend? ______________

2. What number did the shopkeeper add to the subtrahend?

   ____

3. Find the difference: 52 − 17 = ______________

   55 − 20 = ______________

4. The shopkeeper said it was easier to compute 55 − 20. Do you agree or disagree? Explain.

   ____

   ____

   ____

   ____

   ____

   ____

Student Mathematician: ________________________ Date: ____________

Mysterious Mula Subtractions (continued)

At the leather shop Dru started out with 132 mula. Then he bought a new coat for 89 mula. This shopkeeper said, "132 – 89 equals 133 – 90 and 133 – 90 is easier to figure out in my head!"

5. What number did the shopkeeper add to the subtrahend?

____

6. What number did the shopkeeper add to the minuend?

____

7. Find the difference: $132 - 89 =$ ______________

$133 - 90 =$ ______________

8. The shopkeeper said 133 – 90 is easier to subtract. Do you agree or disagree? Explain.

____

____

____

____

____

____

Student Mathematician: ____________________ Date: __________

Shopping at the Bazaar

Pretend you are a shopkeeper in the meerkat bazaar. Rewrite each expression so that the difference remains the same. Make sure your new expression is easy to solve.

1. 87 − 16 = ____________________

2. 72 − 47 = ____________________

3. 123 − 59 = ____________________

4. 624 − 108 = ____________________

5. 512 − 199 = ____________________

6. 281 − 187 = ____________________

7. Choose one of the problems. Explain your choices when you rewrote the expression.

____

____

____

8. Imagine you are a shopkeeper in the bazaar. Make up a word problem and solve it.

____

____

Student Mathematician: ______________________ Date: __________

Equal Expressions Recording Sheet

Use your *Equal Expressions* cards from the game to write "equal" equations.

1. ______________________

2. ______________________

3. ______________________

4. ______________________

5. ______________________

6. ______________________

7. ______________________

8. ______________________

9. ______________________

Student Mathematician: ________________________ Date: __________

THINK DEEPLY

Dru and Teller gave a shopkeeper in the bazaar 32 mula for a basket that cost 19 mula. The shopkeeper needed to figure out the amount of change to give them. Instead of figuring out 32 – 19, the shopkeeper said, "I owe you 33 – 20 mula in change."

a. Do these expressions have the same value? Explain.

____

____

____

b. Why did the shopkeeper change the expression to 33 – 20?

____

____

____

c. Make up another expression that has the same value as 32 – 19. Explain your thinking.

____

____

____

Student Mathematician: ______________________________ Date: ____________

Designing Twice as Nice Belts #1

Color the squares and write your answers on the blanks.

1. Make a belt where the first strand has 15 red stones. The second strand has 5 blue stones and the rest are yellow stones.

   Second strand: ______________ yellow stones

   ______________ blue stones

2. Make a belt where the second strand has 8 red and 7 blue stones. The first strand has the same two colors. There are 3 more red stones than blue stones on the first strand.

   First strand: ______________ red stones

   ______________ blue stones

3. Make a belt where the first strand has 4 fewer red stones than blue stones. The second strand has 6 red and 6 blue stones.

   First strand: ______________ red stones

   ______________ blue stones

Student Mathematician: ______________________________ Date: ____________

Designing Twice as Nice Belts #1 (continued)

Color the squares and write your answers on the blanks.

4. Make a belt where the first strand has 15 red stones. The second strand has twice as many blue stones as yellow stones.

Second strand: ______________ yellow stones

______________ blue stones

5. Make a belt where the second strand has 8 orange and 6 green stones. The first strand has the same two colors. There are 4 more orange stones than green stones on the first strand.

First strand: ______________ orange stones

______________ green stones

6. Make a belt where the first strand has 5 fewer red stones than blue stones. The second strand has 15 stones. 6 are red and the rest are blue stones.

First strand: ______________ red stones

______________ blue stones

Student Mathematician: ________________________ Date: ____________

Designing Twice as Nice Belts #2

Write the number of each color on the bar model.

1. Make a belt with 5 red stones and twice as many blue stones as red on the first strand. The second strand should have 2 fewer red stones than the first strand and the rest blue stones.

5 red	______ blue
_____ red	______ blue

2. Make a belt with 12 green stones and three fewer yellow stones than green on the first strand. The second strand should have 6 green stones and the rest yellow stones.

12 green	_____ yellow
_____ green	_____ yellow

3. Make a belt with 3 red stones and 11 more yellow stones than red on the first strand. The second strand should have 10 red stones and the rest are yellow stones.

3 red	_____ yellow
_____ red	_____ yellow

4. Make a belt with six yellow stones and twice as many green stones as yellow on the first strand. The second strand should have two more yellow stones than the first and the rest green stones.

______ green	______ yellow
______ green	______ yellow

Student Mathematician: ______________________ Date: ____________

Writing and Solving Equations

$r =$ number of red stones $b =$ number of blue stones

$g =$ number of green stones $y =$ number of yellow stones

$o =$ number of orange stones

1. Complete the equation for this bar model.

$g +$ ______________ $= 12$

12	
g	9

There are ______________ green stones.

2. Complete the equation for this bar model.

______________ $= r$

r	
4	11

There are ______________ red stones.

3. Complete the equation for this bar model.

______________ $= o$

o	5
21	

There are ______________ orange stones.

Student Mathematician: ______________________ Date: __________

THINK DEEPLY

The owner of the *Twice as Nice Belt Shop* told Dru and Teller that she is making a belt where the number of green stones on one strand is two more than the number of blue stones on the same strand.

Dru said that they can write the equation $g + 2 = b$ to show this. Teller said that the equation should be $g = 2 + b$.

Who is right? Explain to them using words and a bar model.

____

____

____

____

____

____

____

____

Student Mathematician: ______________________________ Date: __________

Botswana Bazaar Shopping Record Sheet

Name of Shop	Problem Number	Show your work	Answer

Student Mathematician: ______________________ Date: __________

THINK DEEPLY

Dru and Teller gave the *Botswana Basket Shop* owner 500 mula to buy a large basket for 285 mula and a small one for 135 mula.

a. How much do the two baskets cost?

____

b. Dru and Teller get confused when making change. Show how you figured out how much change they should get. Use words, pictures, and computation.

____

____

____

____

____

Student Mathematician: ______________________ Date: __________

Botswana Bazaar Shopping Record Sheet

Name of Shop	Problem Number	Show your work	Answer

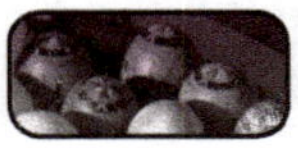

Student Mathematician: ____________________ Date: __________

THINK DEEPLY

Dru and Teller gave the *Botswana Basket Shop* owner 500 mula to buy a large basket for 285 mula and a small one for 135 mula.

a. How much do the two baskets cost?

____

b. Dru and Teller get confused when making change. Show how you figured out how much change they should get. Use words, pictures, and computation.

____

____

____

____

____

Student Mathematician's Glossary

<table>
<tr><td>_______________</td><td>↓ ↓
$2 + 3 = 5$</td></tr>
<tr><td>_______________</td><td><table><tr><td colspan="2">5</td></tr><tr><td>3</td><td>2</td></tr></table></td></tr>
<tr><td>_______________</td><td>$8 + 5 = 5 + 8$</td></tr>
<tr><td>_______________</td><td>the answer to a subtraction problem
↓
$12 - 3 = 9$</td></tr>
<tr><td>_______________</td><td>the symbols used to show a number</td></tr>
<tr><td>_______________</td><td>$=$</td></tr>
</table>

Student Mathematician's Glossary

______________	$5 = 2 + 3$ $5 + 6 = 6 + 5$ $12 - 8 = 4$
______________	6 $7 + 4$ 3×9 $18 - p$
______________	$20 + 0 = 20$ $0 + 20 = 20$
______________	↓ $12 - 3 = 9$
______________	$\neq$

Student Mathematician's Glossary

________________	↓ $12 - 3 = 9$
________________	the answer to an addition problem ↓ $2 + 3 = 5$
________________	a b □

CPSIA information can be obtained
at www.ICGtesting.com
Printed in the USA
LVHW021310110621
689809LV00004B/13

9 781524 910020